Pocket Books

Natural Wonders

Kane Miller
A DIVISION OF EDC PUBLISHING

First American Edition 2016
Kane Miller, A Division of EDC Publishing

Copyright © Green Android Ltd 2016

For information contact:
Kane Miller, A Division of EDC Publishing
P.O. Box 470663
Tulsa, OK 74147-0663
www.kanemiller.com
www.edcpub.com
www.usbornebooksandmore.com

Please note that every effort has been made to check the accuracy of the information contained in this book, and to credit the copyright holders correctly. Green Android Ltd apologize for any unintentional errors or omissions, and would be happy to include revisions to content and/or acknowledgements in subsequent editions of this book.

Printed and bound in China, January 2016
Library of Congress Control Number: 2015947617
ISBN: 978-1-61067-473-7

Images © shutterstock.com: amazon rainforest © 300 river © thobo, angel falls © Richie Ji, antelope canyon, the midnight sun © Filip Fuxa, arizona meteor crater © turtix, atacama desert © Jool-yan, atlas mountains © alex7370, aurora borealis © Gardar Olafsson, balcarka cave © Nadezda Murmakova, bastei rocks © Ugis Riba, bay of fundy ©Josef Hanus, bay of islands, iguazu falls © Lev Kropotov, black sea © Yuri Kravchenko, bryce canyon © Alexey Stiop, buckskin gulch © Michael Zysman, bungle bungle range © Light & Magic Photography, cango caves ©PhotoSky, caspian sea ©ekipaj, castle geysir © DJHolmes86, charyn canyon © Pikuso.kz, cliffs of moher © Kwiatek7, Cotopaxi © Fotos593, dead sea © ChameleonsEye, death valley © tobkatrina, devils tower © Richard A McMillin, dobsimna ice cave © SÃƒÂ¡gi ElemÃƒÂ©r, el tatio geysers © Nataliya Hora, fish river canyon © Pyty, fly geyser © Berzina, fox glacier © VanderWolf Images, fraser island © Antoine Beyeler, galapagos island © SidEcuador, giants causeway © Fulcanelli, glass house mountains © pisaphotography, gobi desert © Lian Deng, grand canyon, el yunque © Jason Patrick Ross, great salt lake © Johnny Adolphson, ha long bay © Igor Plotnikov, horsetail fall © Peggy Sells, huangshan mountain © Stefano Tronci, k2 © Patrick Poendl, komodo island © duchy, lake baikal © Katvic, lake nakuru © Vorobyev Dmitry, lake titicaca © Steve Allen, landscape arch © Zack Frank, les gorges du verdon © sigurcamp, mackenzie river © Max Lindenthaler, marble caves © Marian Dreher, margerie glacier © Nina B, mesoamerican barrier reef © Wollertz, mississippi river © Dan Thornberg, mojave desert © holbox, mont blanc © Venturelli Luca, monument valley © Peter Wey, moraine lake © Richard Cavalleri, mount bromo © phaechin, mount elbrus © Poprotskiy Alexey, mount erebus © Sergey Tarasenko, mount etna © ollirg, mount everest © Anton Rogozin, mount fuji © 10 FACE, mount Kilimanjaro © Graeme Shannon, mount kilimanjaro (p4) © Eduard Kyslynskyy, mount Olympus © Mount Olympus, namib desert © Marisa Estivill, niagara falls © Nikola Bilic, niagra falls (p6) © Javen, paradise bay © photoiconix, parhelic circle © Peter Gudella, perito moreno glacier © kovgabor, phang nga bay © Anna Jedynak, pinnacles national monument © Dreamframer, plitvice lakes © LeonP, puerto-princesa © r.nagy, pulpit rock © Andrey Armyagov, raja ampat reef, new caledonia reef © Ethan Daniels, red sea © helza, red sea coral reef © fotomaton, redwood national park © welcomia, rock of Gibraltar © sokolovsky, sahara desert © Shanti Hesse, seljalandsfoss waterfall © TTstudio, seven sisters © Matt Gibson, shilin stone forestx2 © Mau Horng, smoo cave © Anton Kossmann, solar eclipse © Igor Zh., spotted lake © Rowdy Soetisna, sugarloaf mountain © Donatas Dabravolskas, table mountain © Denis Mironov, teide © Mikadun, the black forest © Andreas Zerndl, the blue cave of bisevo © paul prescott, the great barrier reef © deb22, the great geyser © Sergey Didenko, the matterhorn © elxeneize, the niger © Dutourdumonde Photography, the nile © bumihills, the wave (p1) © kojihirano, thor's hammer © Pierre Leclerc, twelve apostle © Photodigitaal.nl, uluru © Stanislav Fosenbauer, Vesuvius © Ewais, victoria falls © Przemyslaw Skibinski, wadi rum © Radek Sturgolewski, wave rock © Totajla, white desert, salar de uyuni © sunsinger, wolfe creek impact crater © David PETIT, zhangjiajie national forest park © minddream. Images © istock.com: the midnight sun (p7) © Koonyongyut. Images © istock.com: lonar crater © Aditya Laghate, pinguialuit crater © NASA. Courtesy of Denis Sarrazin, fraser island © Adbar, sargasso sea © Mats Halldin, light pillars © Christoph Geisler. Images © www.flickr.com: Amguid Crater © Alan Roberts, christmas island © Kirsty Faulkner.

Charyn Canyon

Factfile

Location	Europe – Almaty, Kazakhstan
Rock type	Volcanic basalt and red gravel
Formed by	Water erosion
Length	96 miles
Width	262 feet (maximum)
Depth	490–980 feet (maximum)

Feature The Charyn Canyon has many gorges cut into multicolored striped rocks. This canyon has been called the "Grand Canyon's little brother."

Fact For a one mile stretch, known as the Valley of Castles, off the Charyn River, there are crumbling pillars edging the floor of the canyon.

Site status	National park

9

Les Gorges du Verdon

Natural wonders

Land marvels

Canyons

Factfile

Location	Europe – France
Rock type	Limestone
Formed by	Water erosion
Length	15 miles
Width	20–328 feet (bottom); 650–4,921 feet (top)
Depth	2,296 feet (maximum)

Feature The Verdon River's turquoise color is caused by its glacial head and minerals in the water. The gorge attracts rock climbers, hikers and canoers.

Fact There are five man-made lakes, or reservoirs, in Les Gorges du Verdon. They were created when five dams were built to contain the water.

Site status	Regional park

Antelope Canyon

Factfile

Location	North America – Arizona, U.S.A.
Rock type	Sandstone
Formed by	Water
Length	600 feet (The Crack); 1,335 feet (The Corkscrew)
Width	3–30 feet
Depth	120 feet (maximum)

Feature This slot canyon is made up of two sections – The Crack and The Corkscrew, which has to be accessed by descending metal stairways.

Fact Sunlight plays upon the canyon's smooth, flowing orange walls. At certain times a single, strong beam of light will shoot into the canyon.

Site status	Regional park

Bryce Canyon

Factfile

Location	North America – Utah, USA
Rock type	Limestone, siltstone, dolomite and mudstone
Formed by	Water / frost wedging
Length	20 miles
Width	3 miles (maximum)
Depth	800 feet (maximum)

Feature For thousands of years, wind erosion has been shaping the rocks in the canyon into tall, thin pillars, known as hoodoos. They can be up to 100 feet tall.

Fact Though called a canyon, it is a series of vast amphitheaters of red, orange, pink and white rock eroded into arches, windows, bridges and pillars.

Site status   **National park**

Buckskin Gulch

Factfile

Location	North America – Utah and Arizona, USA
Rock type	Sandstone
Formed by	Water
Length	15 miles
Width	10–20 feet
Depth	500 feet

Feature With its striking red Navajo sandstone, Buckskin Gulch is one of the world's longest and deepest slot canyons. The sun rarely reaches the gulch's floor.

Fact With an average of eight flash floods a year, climbers have to exercise extreme caution. It is regarded as one of the top 10 most dangerous US hikes.

Site status	**National monument**

Grand Canyon

Factfile

Location	North America – Arizona, USA
Rock type	Limestone, shale, sandstone and composite rocks
Formed by	Water erosion
Length	277 miles
Width	600 feet–18 miles
Depth	1 mile (maximum)

Feature This canyon was eroded over two billion years by the Colorado River and its tributaries. There are over 100 named rapids in the canyon.

Fact The Grand Canyon was discovered by a Spaniard, Garcia Lopez de Cardenas in 1540, but Native North American tribes lived here from 1200 BCE.

Site status — World Heritage Site

14

Cango Caves

Factfile

Location	Africa – Western Cape, South Africa
Rock type	Limestone
Length	2.5 miles
Chamber size	321 feet long (largest chamber)
Access	By foot
Known for	Cleopatra's Needle

Feature Enormous caverns, one as long as a soccer field, contain dripstone formations like stalagmites, stalactites and antler-like helictites.

Fact Stalagmites grow from the ground up. Cleopatra's Needle is a 30-foot-high orange-and-gold colored stalagmite that is over 150,000 years old.

Site status	National monument

15

Puerto-Princesa

Factfile

Location	Asia - Palawan Province, Philippines
Rock type	Limestone
Length	5 miles
Chamber size	1,181 feet long (largest chamber)
Access	By boat
Known for	Limestone karst landscape, underground river

Feature An opening in a cliff leads to a five-mile-long river set five miles under Mt. St. Paul. The Italian's Chamber is one of the largest caves in the world.

Fact Puerto-Princesa, the longest navigable underground river in the world, flows from a lagoon, under a mountain and into the South China Sea.

Site status   World Heritage Site

Balcarka Cave

Factfile

Location	Europe – Czech Republic
Rock type	Limestone
Length	3,773 feet
Chamber size	213 feet by 66 feet by 49 feet (largest chamber)
Access	By foot
Known for	Stalactite and stalagmite formations

Feature Stalagmites rise from the cave floor and stalactites hang from the ceiling among enormous columns in this colorful labyrinth cave system.

Fact The Balcarka Caves were once lived in by humans. Evidence has been found of Early Stone Age people, and even items from medieval times.

Site status	**Nature reserve**

17

Dobsina Ice Cave

Factfile

Location   Europe – Slovakia
Rock type   Limestone and ice
Length   4,892 feet
Chamber size   396 feet by 148 feet by 39 feet (largest chamber)
Access   By foot
Known for   Being used for skating

Feature Entry to the amazing rock caves filled with ice stalagmites and stalactites is through an ice hole, followed by a descent along icy tunnels.

Fact Created when a glacier flooded the caves, the glacial ice is moving through the cave system at the slow rate of 0.8–1.6 inches each year.

Site status	World Heritage Site

Smoo Cave

Factfile

Location   Europe – Scotland
Rock type   Limestone
Length   272 feet (accessible)
Chamber size 200 feet by 130 feet by 50 feet (largest chamber)
Access   By foot and boat
Known for   Its dramatic setting

Feature The entrance to Smoo Cave is the largest sea cave entrance in Great Britain. It has been likened to walking into an enormous opened mouth.

Fact The sea carved the first chamber and a river (burn) shaped the second chamber, which features a waterfall dropping 80 feet into a deep pool.

Site status	Conservation area

Marble Caves

Factfile

Location   South America – Chile
Rock type   Marble
No. of caves   3
Age   6,200 years
Access   By boat
Known for   Stunning blue and gray chambers

Feature The marble reflects the azure-colored water of the lake. In spring the caves are turquoise, and in summer the water rises and the walls turn a deep blue.

Fact The caves, also called the Marble Cathedral, include caverns, pillars and tunnels that were formed by wave action over a period of 6,200 years.

Site status   **National monument**

Amguid Crater

Factfile

Location	Africa – Algeria
Diameter	1,640–1,740 feet
Depth	165 feet
Age	10,000–100,000 years old
Caused by	Meteorite impact
Known for	Being the best preserved crater on Earth

Feature This impact crater is almost perfectly circular with steep walls. The raised rim of the crater is edged with sandstone rocks several yards across.

Fact Amguid is rarely visited. It is located in a remote part of the Sahara, access is only by foot and the nearest settlement is about 60 miles away.

Site status   **Geographical landmark**

Lonar Crater

Factfile

Location	Asia – Lonar, India
Diameter	5,900 feet (crater), 3,900 feet (lake)
Depth	490 feet
Age	52,000–570,000 years old
Caused by	Meteorite impact
Known for	Being the world's oldest crater and third largest

Feature The crater is filled with water that is saline around the edge and alkaline within. The lake's microorganisms are rarely encountered elsewhere.

Fact Scientists think the crater was made by a meteorite weighing more than two million tons. It crashed into Earth at a speed of 55,900 miles per hour.

Site status   **Geographical landmark**

Wolfe Creek Impact Crater

Factfile

Location	Australia – Kimberley region, Western Australia
Diameter	2,887 feet
Depth	393 feet (original), 164 feet (current)
Age	300,000 years old
Caused by	Meteorite impact
Known for	The Aboriginal dreamtime story of its creation

Feature When the meteorite crashed, debris was forced up and out, and a round crater was formed. It is the world's second-largest rimmed crater.

Fact Iron meteorite pieces can be found around and in the crater but also miles away. Those in the crater look like rusty balls fused to rocks.

Site status	National park

Arizona Meteor Crater

Natural wonders

Land marvels

Craters

Factfile

Location	North America – Arizona, USA
Diameter	4,100 feet
Depth	570 feet
Age	50,000 years old
Caused by	Meteorite impact
Known for	Being privately owned by the Barringer family

Feature A huge bowl-shaped pit in the ground, the Arizona Meteor Crater is famous for being the best-preserved meteorite impact site on Earth.

Fact The crater was made by a 130-foot-wide asteroid moving at 7.5 miles per second. The impact explosion was equal to 2.5 million tons of TNT.

Site status   **National natural landmark**

24

Pingualuit Crater

Factfile

Location	North America – Quebec, Canada
Diameter	11,286 feet
Depth	875 feet (lake), 1,312 feet (crater)
Age	1.4 million years
Caused by	Meteorite impact
Known for	The lake's exceptionally clear, blue water

Feature Within this crater is one of the deepest lakes in North America. The lake, which fills only with rain and snow, contains some of the world's purest fresh water.

Fact The crater's rim rises 520 feet above the surrounding tundra. The rim was probably higher in the past, but it has been slowly eroded by glaciers.

Site status	National park

25

Namib Desert

Natural wonders

Land marvels

Deserts

Factfile

Location	Africa – Namibia, Angola and South Africa
Environment	Cold coastal desert
Elevation	Sea level–3,000 feet
Temperature	50 °F (coastal region) / 82 °F (inland)
Area	52,000 square miles
Precipitation	0.5–2.0 inches some years; regular dew and fog

Feature Nicknamed the "Skeleton Coast," the Namib's dunes are littered with shipwrecks. The coast is mined for its large deposits of alluvial diamonds.

Fact A cold desert gets moisture from fog or snow, not rain. Fog in the northern and central sections extend 18 miles or more inland from the coast.

Site status   **National park**

Sahara Desert

Factfile

Location	Africa – Algeria, Chad, Egypt, Libya, Mali, Mauritania, Morocco, Niger, Sudan and Tunisia
Environment	Hot desert
Elevation	436 feet below sea level–11,204 feet
Temperature	122 °F (high) / 5°F (low)
Area	3,500,000 square miles
Precipitation	Below 1 inch–4 inches per year

Feature The landscapes of the Sahara include dunes (up to 590 feet high), sand seas, plateaus of stone, gravel and salt flats, volcanoes and mountain ranges.

Fact The Sahara is the world's largest hot desert. Hot deserts receive less than 10 inches of rain a year. It occupies 10 percent of the African continent.

Site status	World Heritage Site

27

White Desert

Factfile

Location	Africa – Farafra, Western Egypt
Environment	Hot desert
Elevation	300 feet
Temperature	113°F (high) / 32°F (low)
Area	116 square miles
Precipitation	0 – 1 inch per year

Feature White chalk rocks standing like statues in this desert were carved by sandstorms over hundreds of years into strange and gravity-defying shapes.

Fact The White Desert was once a seabed where compressed sediment and dead fauna formed chalk. When the sea dried, the chalk was exposed to the elements.

Site status   National park

Gobi Desert

Factfile

Location	Asia – southern Mongolia and northern China
Environment	Cold desert
Elevation	5,000 feet (maximum)
Temperature	122 °F (high) / -40 °F (low)
Area	500,000 square miles
Precipitation	2–7.6 inches per year

Feature It was in the Gobi Desert that the first fossilized dinosaur egg was discovered, but this rock desert also has ice-filled canyons and flaming-red cliffs.

Fact Asia's largest desert is growing! To stop this desertification, China is planting trees along the desert's edge to make a "Green Wall of China."

Site status	Nature reserve

29

Wadi Rum

Factfile

Location	Asia – southern Jordan
Environment	Hot desert
Elevation	5,689 feet
Temperature	104 °F (high) / 39 °F (low)
Area	280 square miles
Precipitation	2–4 inches per year

Feature Rock carvings (petroglyphs) at Wadi Rum show that it has been inhabited for over 12,000 years. One carving is believed to be the first-ever alphabet.

Fact Extremely hot in summer and freezing in winter, Wadi Rum, or the "Valley of the Moon" as it is known, features gorges, arches, caves and high cliffs.

Site status   World Heritage Site

Death Valley

Factfile

Location	North America – eastern California, USA
Environment	Hot desert
Elevation	282 feet below sea level–11,049 feet
Temperature	134 °F (high) / 15 °F (low)
Area	5,156 square miles
Precipitation	2.4 inches per year (maximum)

Feature Death Valley is the largest national park in the US and its sand dunes, salt pan, oases and canyons welcome almost a milllon visitors a year.

Fact Death Valley is the hottest and driest area in the US. In July 1913 the temperature reached 134 °F – the highest recorded anywhere in the world!

Site status   **National park**

31

Mojave Desert

Natural wonders

Land marvels

Deserts

Factfile

Location North America – California, Utah, Nevada and Arizona, USA

Environment Hot desert

Elevation 3,000–6,000 feet

Temperature 105 °F (high) / 8 °F (low)

Area 25,000 square miles

Precipitation 2–4 inches per year

Feature Surrounded by mountains, the Mojave is a sandy desert with few plants and white salt flats left behind when rainwater evaporates.

Fact Under the Mojave's unwelcoming surface is where its treasures lay. People have mined the desert for years for salt, silver, gold and iron.

Site status	Nature reserve

32

Painted Desert

Factfile

Location	North America – northern Arizona, USA
Environment	Cold desert
Elevation	4,500–6,500 feet
Temperature	105°F (high) / -25°F (low)
Area	146 square miles
Rainfall	5–9 inches per year

Feature This barren badlands is a rainbow in rock. Red, orange, pink, white, mauve and gray minerals have colored the layers of shale and mudstone.

Fact This desert, which lies partially in the Navajo Nation, was formed by many geological events – volcanic eruptions, floods and earthquakes – over millennia.

Site status	National park

Atacama Desert

Factfile

Location	South America – northern Chile
Environment	Cold desert
Elevation	3,000–5,000 feet
Temperature	81°F (high) / 28°F (low)
Area	54,000 square miles
Rainfall	0.06 inches per year (average)

Feature With red rocky canyons and salt flats, the Atacama Desert has a landscape like the planet Mars, which is why NASA test their Mars Rovers here!

Fact Areas of this desert have no records of rainfall and it is likely that no rain fell in the Atacama from 1570–1971! It is the driest non-polar region on Earth.

Site status	National park

34

The Black Forest

Factfile

Location	Europe – Baden-Württemberg, Germany
Environment	Forests, mountains, wetlands, springs and rivers
Area	2,320 square miles
Features	Source of Danube and Neckar rivers
Plant species	Around 1,980
Rainfall	18 inches per year

Feature In the southwest of Germany, the Black Forest is famous for its mountains, waterfalls and for the fairy-tale villages that are hidden among its trees.

Fact The forest is also famous for Black Forest gateaux, a chocolate cake, and for cuckoo clocks, which have been made in this area since the 1700s.

Site status **National park**

35

El Yunque

Natural wonders

Land wonders

Forests

Factfile

Location	North America – Puerto Rico, Caribbean
Environment	Tropical vegetation, rivers and waterfalls
Area	45 square miles
Features	High rainfall, carvings and a dwarf forest
Plant species	240 of which 23 are unique
Rainfall	240 inches per year

Feature This rain forest is usually blanketed under thick white clouds. The dwarf forest of short, fat trees with few leaves is due to the acidity of the soil.

Fact Many granite boulders near the rivers and streams display carvings (petroglyphs) done by the Taino people hundreds of years ago.

Site status   **National forest**

Redwood National Park

Factfile

Location	North America – California, USA
Environment	Woodlands, rivers, prairies and coast
Area	206 square miles
Features	Giant trees and abundant marine and land fauna
Plant species	856
Rainfall	60–80 inches

Feature The General Sherman is the biggest tree in the world – 274.9 feet high and a girth of 102.6 feet at ground level gives it the largest trunk volume of all!

Fact Trees in this forest are among Earth's oldest and tallest living things. Some sequoias are 2,000 years old, and the record redwood was 379.1 feet tall!

Site status **World Heritage Site**

Amazon Rain Forest

Natural wonders

Land wonders

Forests

Factfile

Location	South America – Brazil, Bolivia, Peru, Ecuador, Colombia, Venezuela, Guyana, Suriname and French Guiana
Environment	Savannas, flood plain forests and swamps
Area	2.3 million square miles
Features	A diverse ecosystem of flora and fauna
Plant species	Over 40,000
Rainfall	120 inches per year

Feature The largest rain forest in the world, covering 40 percent of South America, is fed by the Amazon, the world's second-longest river that has 1,100 tributaries.

Fact As the trees here have dense canopies of branches and large leaves, it can take nearly 10 minutes for a raindrop to reach the rain forest floor.

Site status	World Heritage Site

Christmas Island

Factfile

Location	Indian Ocean – 1,560 miles from Australia
Environment	Beaches, cliffs, caves and tropical rain forest
Area	52 square miles
Population	1,530
Elevation	Sea level–1,184 feet
Known for	Annual migration of 100 million red crabs

Feature The island was once the summit of a volcano. It is host to tropical rain forests that boast 135 plant species including 18 found nowhere else on Earth.

Fact Christmas Island is an isolated Australian territory. Its range of native seabirds and crabs has earned it the name "Galapagos of the Indian Ocean."

Site status   **National park**

39

Komodo Island

Natural wonders

Land wonders

Islands

Factfile

Environment   Asia – Indonesia
Location   Rugged hills, beaches and cliffs
Area   131 square miles (Komodo Island only)
Population   2,000
Elevation   2,400 feet
Known for   Its unique Komodo dragons and pink beach

Feature Komodo National Park includes three islands, and their reefs, mangroves and seagrass beds contain 1,000 fish species and many marine mammals.

Fact The island is famous for its population of 5,700 Komodo dragons. At 10 feet long and weighing 300 pounds, they are the world's largest lizards.

Site status   World Heritage Site

Fraser Island

Factfile

Environment	Australia – off the Queensland coast
Location	Lakes, dunes, tropical rain forests and fauna
Area	710 square miles
Population	Approximately 200
Elevation	787 feet
Known for	Being the largest sand island in the world

Feature Fraser Island with its unique rain forests growing on sand has 150 miles of white beaches, colored sand cliffs and over 100 freshwater lakes.

Fact Formed over 750,000 years by the buildup of sand trapped on top of a volcanic bedrock, its sand contains a special fungi that gives nutrients to plants.

Site status   **World Heritage Site**

Galapagos Island

Natural wonders

Land wonders

Islands

Factfile

Environment	Pacific Ocean – 620 miles west of South America
Location	Shield volcanoes, coastal cliffs and arid areas
Area	29,595 square miles (UNESCO estimate)
Population	26,000
Elevation	88 °F (high) / 66 °F (low)
Known for	Its preserved and intact natural ecosystem

Feature The wildlife of the 19 volcanic islands, like the land iguana, giant tortoise and finches, and the marine reserve, are unique to this isolated archipelago.

Fact The Galapagos Islands were visited and studied by Charles Darwin in 1835, and have been called "a living laboratory of evolutionary change."

Site status   **World Heritage Site**

Atlas Mountains

Factfile

Location	Africa – Morocco, Algeria and Tunisia
Type	Series of mountain ranges
Highest point	13,671 feet
Range length	1,553 miles
Temperature	84 °F (high) / 55 °F (low)
Known for	Being home to the endangered Barbary leopard

Feature The Tizi n'Test is a high, narrow pass through the Atlas. Its 2,092 miles of hairpin corners takes in walled cities, ancient sites and spectacular scenery.

Fact The Atlas run from the Atlantic Ocean to the Mediterranean Sea along the western edge of the Sahara Desert. They are rich in metals, coal and natural gas.

Site status   **National park**

43

Mount Kilimanjaro

Natural wonders

Land wonders

Mountains and volcanoes

Factfile

Location	Africa – northeastern Tanzania
Type	Dormant volcanic mountain with three cones
Highest point	19,341 feet
Last eruption	150,000–200,000 years ago
Temperature	-17 °F (summit) / 70 °F (base)
Known for	Being a snowcapped mountain on the equator

Feature Kilimanjaro is the highest freestanding mountain in the world. Popular with climbers, the 5 to 9-day ascent goes from tropical to Arctic ecosystems.

Fact Kilimanjaro's name is a mystery. Does it mean mountain of light, caravans or greatness? The summit is named Uhuru, meaning "freedom peak."

Site status   World Heritage Site

Table Mountain

Factfile

Location	Africa – Cape Town, South Africa
Type	Flat-topped, sandstone mountain
Highest point	3,563 feet
Range length	Part of the 37-mile-long Cape Fold range
Temperature	72 °F (high) / 51 °F (low) at the summit
Known for	Its level plateau and unique flowering shrubland

Feature To reach the mountain's two-mile-long plateau, there are 900 hiking and climbing routes and a cable car that does the trip in five minutes.

Fact The southeasterly Cape Doctor winds cause clouds, called "The Tablecloth" to form over the mountain. The winds are said to blow away pestilence.

Site status **World Heritage Site**

Mount Erebus

Factfile

Location   Antarctica – Ross Island
Type   Active shield/conical volcano with one cone
Highest point 12,448 feet
Last eruption December 1972 (ongoing)
Temperature   -4 °F (high) / -58 °F (low)
Known for   Being Earth's most southern active volcano

Feature Erebus is covered by glaciers and its crater contains a 328-foot-deep lake of boiling lava. The island is also home to 500,000 Adelie penguins.

Fact Mount Erebus Volcanic Observatory has a webcam on the volcano's rim. The solar and wind powered camera sends live images of the lava lake.

Site status	National park

Huangshan Mountain

Factfile

Location   Asia – southern Anhui province, eastern China
Type   Granite-peaked mountain range
Highest point 6,115 feet
Range length   160 miles
Temperature   91 °F (high) / 28 °F (low)
Known for   Being the most beautiful mountain in China

Feature Over 60,000 steps are cut into the rocks of Huangshan, which means yellow mountains. It was named for a mythical ancestor, the Yellow Emperor.

Fact A Chinese legend has it that Huangshan was the source of an elixir of life. Artists and poets have been inspired by its trees and peaks in a sea of clouds.

Site status   **World Heritage Site**

47

K2

Factfile

Location	Asia – Karakoram Range, China–Pakistan border
Type	Isolated six-sided pyramid of gneissic rock
Highest point	28,251 feet
Range length	311 miles
Temperature	104 °F (maximum at base) / -40 °F (summit)
Known for	Being the second-highest mountain in the world

Feature Unpredictable storms, steep icy slopes, high altitude and being in the most glaciated region outside of the poles makes K2 unclimbable during winter.

Fact Known as the "Savage Mountain," 335 people have made the climb, but of those 82 have died. It is the most difficult mountain to climb.

Site status   **National park**

Mount Bromo

Factfile

Location   Asia – East Java, Indonesia
Type   Highly active cone inside a giant caldera
Highest point 7,641 feet
Last erupted   December 2010–January 2011
Temperature   68 °F (high) / 37.4 °F (low) at the summit
Known for   Its sunrise views and awe-inspiring scenery

Feature Locals and visitors flock to the area to witness the sun rising over Mount Bromo, the Sea of Sand, Semeru volcano and the Tengger massif (range).

Fact During the month-long Yadnya Kasada festival, gifts of food are thrown into this volcano, which is named after Brahma, the Hindu creator god.

Site status   **National park**

49

Mount Elbrus

Natural wonders

Land marvels

Mountains and volcanoes

Factfile

Location   Asia and Europe – Caucasus Mountains, Russia
Type   Dormant volcano filled with ice and snow
Highest point   18,510 feet
Last eruption   50 CE
Temperature   5 °F (high) / -22 °F (low) at the summit
Known for   Having two summits (both dormant domes)

Feature The bathroom at the summit is the highest in Europe and also the scariest. It is covered in ice and is precariously perched on the end of a rock!

Fact Mount Elbrus is the highest mountain peak in Europe and attracts many climbers as the ascent is easy, but bad weather has caused many fatalities.

Site status   **National park**

Mount Everest

Factfile

Location	Asia – Himalaya Mountains, Nepal/Tibet
Type	Shale, limestone and marble mountain
Highest point	29,035 feet
Range length	1,550 miles
Temperature	-2.2 °F (high) / -32.8 °F (low) summit averages
Known for	Being the highest point on Earth

Feature Even though the peak is snow-covered year round and low in oxygen, a few animals survive here. One is a jumping spider that lives at 22,000 feet!

Fact Everest is called "Chomolunga" in Tibet, which means goddess of the universe. It grows by 0.15 inches a year as India's landmass moves.

Site status   **World Heritage Site**

Mount Fuji

Natural wonders

Land marvels

Mountains and volcanoes

Factfile

Location	Asia – Honshu Island, Japan
Type	Active volcano
Highest point	12,388 feet
Last eruption	December 1707
Temperature	40.8 °F–43.6 °F (summit average in July–August)
Known for	Being one of Japan's three holy mountains

Feature Because of its sacred status, women were banned from the summit until the 1880s. The samurai did their training at the base of Mount Fuji.

Fact The symmetrical cone of Mount Fuji can be seen on clear days from the capital Tokyo. It is a place of pilgrimage and has inspired artists for centuries.

Site status | **World Heritage Site**

52

Glass House Mountains

Factfile

Location   Australia – Sunshine Coast, Queensland
Type   Volcanic plugs exposed after erosion
Highest point 1,821 feet
Site area   3.5 square miles
Temperature   83 °F (average high) / 61 °F (average low)
Known for   Its collection of dramatic sheer-sided peaks

Feature Thirteen peaks, made from cooled lava, rise up from the coastal plain and glow in the sunlight. Captain Cook in 1770 said they looked like fiery glass furnaces.

Fact Sacred to local Aboriginals, there is a legend that each mountain represents a member of a family and the story of that family.

Site status	National park

Mont Blanc

Natural wonders

Land marvels

Mountains and volcanoes

Factfile

Location	Europe – the Alps, Italian-French border
Type	Crystalline rocks carved by glaciers
Highest point	15,770–15,782 feet
Range length	25 miles
Temperature	23 °F (summit high) / -45.4 °F (summit lowest)
Known for	The hot tub that was built on the summit in 2007

Feature Mont Blanc, or White Mountain, is domed in ice and snow, and 40 square miles are covered by glaciers. A 7.25-mile tunnel runs under the mountain.

Fact Mont Blanc is Europe's second-highest mountain and attracts 20,000 climbers a year. It is the birthplace of modern mountaineering.

Site status   **National park**

54

Mount Etna

Factfile

Location	Europe – Sicily, Italy
Type	Active volcano with explosions and lava flows
Highest point	10,991 feet (currently)
Last eruption	May 2015 but constantly active
Temperature	68 °F (average high) / 28 °F (average low)
Known for	Being Europe's tallest and most active volcano

Feature Etna has been active for over 2.6 million years, and continues to erupt to this day. More than a quarter of Sicily's population live on its rich, fertile slopes.

Fact It is impossible to fix the height of Mount Etna as it can grow or shrink with each eruption. This volcano erupts smoke rings, which is a very rare event.

Site status	World Heritage Site

Mount Olympus

Natural wonders

Land marvels

Mountains and volcanoes

Factfile

Location	Europe – near Thessaloniki, Greece
Type	Limestone with granite, carved by glaciers
Highest point	9,573 feet
Site area	63 square miles
Temperature	68 °F (high) / -4 °F (low) at summit
Known for	Home of the 12 Greek gods and throne of Zeus

Feature In 1981 UNESCO declared areas of the Mount Olympus National Park a biosphere reserve to protect them from grazing and timber felling.

Fact The highest mountain in Greece, Olympus is one of 52 peaks in the massif, which contain ravines, woodlands, grasslands and alpine fields.

Site status	**Nature reserve**

Teide

Factfile

Location   Europe – Tenerife, Canary Islands, Spain

Type   Active conical volcano on top of a shield volcano

Highest point 12,198 feet

Last eruption 1909

Temperature   72.5 °F (high) / 33.8 °F (low)

Known for   The world's third-tallest volcanic structure

Feature The native Guanches believed the devil was trapped inside the volcano and escaped during eruptions. Teide was also thought to hold up the sky.

Fact Mount Teide is one of many volcanoes in the Las Cañadas caldera, a 10-mile-wide circle of mountains that were created by an older volcano.

Site status   **World Heritage Site**

57

The Matterhorn

Natural wonders

Land marvels

Mountains and volcanoes

Factfile

Location	Europe – French-Swiss-Italian border
Type	Pyramid-shaped peak of gneiss, cut by glaciers
Highest point	14,692 feet
Range length	25 miles
Temperature	39 °F (high) / -14.5 °F (low) at summit
Known for	Being the last great Alpine peak to be climbed

Feature The Matterhorn has two summits: an eastern Swiss one, and a western Italian one, that are connected by a 330-foot-long rocky ridge.

Fact The four faces, especially the north face, of this peak are extremely steep (little ice or snow clings to them) and avalanches are frequent.

Site status	National park

58

Vesuvius

Factfile

Location	Europe – near Naples, Italy
Type	Active volcano with explosions and lava flows
Highest peak	4,203 feet
Last eruption	March 1944
Temperature	85 °F (high) / 39 °F (low)
Known for	Producing the largest eruptions in Europe

Feature Vesuvius overlooks a city of 3 million people and sits in the crater of an older volcano. It is one of several volcanoes, some under the sea, in a volcanic arc.

Fact At 200,000 years old, Vesuvius is a young volcano. It had lain dormant for years until 79 CE when it erupted, burying Pompeii and Herculaneum.

Site status   **National park**

59

Cotopaxi

Factfile

Location	South America – Andes Mountains, Ecuador
Type	Active conical volcano with lava flows
Highest point	19,393 feet
Last erupted	August 2015
Temperature	57 °F (high) / 37 °F (low)
Known for	Being a volcano in the Pacific Rim of Fire

Feature Cotopaxi, once known as "Rain Sender" to the Andeans, is a dangerous volcano. The area is scarred by mud flows, rocks and ash.

Fact Cotopaxi's last major eruption was in August 2015 when it shot hot ash seven miles up into the air. Prior to this it had been dormant for 70 years.

Site status	National park

Sugarloaf Mountain

Factfile

Location   South America – Rio de Janeiro, Brazil
Type   Monolithic granite and quartz peak
Highest point 1,299 feet
Site area   0.8 square miles (estimate)
Temperature   86 °F (high) / 64 °F (low)
Known for   Rising straight from the water's edge

Feature A cable car ride up Sugarloaf Mountain is the best way to see Rio de Janeiro. The trip takes just three minutes. Sugarloaf is popular with rock climbers.

Fact Sugarloaf gets its name from the conical molds that were used to shape large blocks of loaf sugar for shipping in the 16th century.

Site status	World Heritage Site

61

Shilin (Stone Forest)

Natural wonders

Land marvels

Rocks

Factfile

Location	Asia – Yunnan Province, China
Part of	Shilin National Scenic Area
Size	186 square miles
Age	270 million years
Rock types	Limestone and dolomite on sandstone
Known for	Oddly shaped formations and pillars of stone

Feature Shilin's hidden rivers, caves and pillars were carved by water, wind and earthquakes. The Ashima pillar is said to be a runaway girl who was turned to stone.

Fact Between August and November each year, strong gales tear out of one Shilin cave every 30 minutes, giving it the name, Strange Wind Cave.

Site status — World Heritage Site

Zhangjiajie Forest

Factfile

Location	Asia – Hunan Province, China
Part of	Zhangjiajie Sandstone Peak Forest National Geopark
Size	1,400 square miles
Age	380 million years (estimate)
Rock type	Quartz and sandstone
Known for	Towering sandstone columns

Feature The pillars were formed by the action of expanding ice in winter and the effect of plants growing on the rocks. Over 1,000 towers are over 650 feet tall.

Fact The Hallelujah Mountain in the film, *Avatar*, was based on Zhangjiajie's "South Pillar of the Heaven." The real pillar has been renamed in its honor.

Site status World Heritage Site

Bungle Bungle Range

Cities of the world

Land marvels

Rocks

Factfile

Location	Australia – northern Western Australia
Part of	Purnululu National Park
Size	174 square miles (Bungle Bungle only)
Age	350 million years
Rock type	Sandstone
Known for	Its orange-and-black horizontally striped domes

Feature Cathedral Gorge in the park has fantastic natural acoustics and has been used for concerts. In the wet season a pool forms in this amphitheater of red rock.

Fact Resembling giant beehives, the rocks and gorges at Bungle Bungle were formed from layers of clay sediment in an old river bed.

Site status **World Heritage Site**

Twelve Apostles

Factfile

Location	Australia – Port Campbell, Victoria
Part of	Twelve Apostles Marine National Park
Height	150 feet (maximum)
Age	6,000 years
Rock type	Limestone
Known for	There being only eight apostles still standing

Feature The 10-mile marine park is not just about the beauty above water. Underwater there are sloping reefs, arches, canyons, fissures and gutters.

Fact The pounding of the Southern Ocean first eroded caves in the cliffs, and then eroded the cave roofs. When these fell in, the Apostles were created.

Site status	National park

Uluru

Factfile

Location	Australia – Northern Territory
Part of	Uluru–Kata Tjuta National Park
Size	1,141 feet high, 2.2 miles long and 1.2 miles wide
Age	600–700 million years
Rock type	Sandstone
Known for	Changing color from sunrise through to sunset

Feature This famous "island mountain" extends underground to a depth of 1.5 miles. It is noted for being just one type of rock, and lacking fractures.

Fact Uluru is sacred to the Anangu people, the traditional custodians of the land around Uluru. The Anangu own the national park and lead the walking tours.

Site status	World Heritage Site

Wave Rock

Factfile

Location	Australia – Hyden, Western Australia
Part of	Hyden Wildlife Park
Size	46 feet high and 360 feet long
Age	2.7 billion years
Rock type	Granite
Known for	Its resemblance to a tall, breaking wave

Feature During the wet season, spring water runs down The Wave and reacts with the granite, leaving red, yellow and brown streaks down the rock face.

Fact The ancient granite has been weathered by wind and water for 60 million years, eroding the base but leaving an overhang at the top.

Site status	Nature reserve

Bastei Rocks

Factfile

Location	Europe – Rathan, Germany
Part of	Saxon Switzerland National Park
Size	636 feet high (maximum above the river)
Age	1 million years
Rock type	Sandstone
Known for	Stunning views from the Bastei Bridge

Feature The sandstone Bastei Bridge that links the jagged peaks and offers views of the Elbe River is 262 feet long and hundreds of feet above the valley floor.

Fact Bastei Rocks, also called the "City of Stone," has been a tourist site for over 200 years. Initially access was via a steep staircase of 487 steps!

Site status	National park

Cliffs of Moher

Factfile

Location	Europe – County Clare, Ireland
Part of	Burren and Cliffs of Moher Geopark
Size	702 feet high (maximum) and 5 miles long
Age	300-million-year-old rocks
Rock type	Shale, siltstone, limestone and sandstone
Known for	The sheer cliffs and distant views

Feature The stormy Atlantic Ocean pounds the cliffs, exposing the 300-million-year-old sediment and trace fossils that were compressed to form the rocks.

Fact This wall of sheer cliffs is a natural star *and* a film star. They were used as a location in *Harry Potter and the Half-Blood Prince*.

Site status   **National park**

Giant's Causeway

Factfile

Location	Europe – County Antrim, Northern Ireland
Part of	Giant's Causeway and Causeway Coast
Size	0.3 square miles
Age	50–60 million years
Rock type	Basalt
Known for	40,000 polygonal columns up to 82 feet long

Feature The columns were formed during volcanic activity when a plateau of molten basalt cooled, contracted and cracked to form multisided pillars.

Fact Legend has it that the causeway was made by a giant called Finn MacCool, so that he could fight a giant who lived on the Scottish island of Staffa.

Site status **World Heritage Site**

Pulpit Rock

Factfile

Location	Europe – Forsand, Norway
Part of	National Tourist Route (Ryfylke)
Size	82 feet long by 82 feet wide
Age	10,000 years
Rock type	Granite
Known for	Being one of the world's top 10 viewing points

Feature Sitting 2,000 feet above the fjord, this flat outcrop is a magnet to hikers. Existing cracks in Pulpit Rock mean that it will fall down in the future.

Fact When water from an Ice Age glacier froze and expanded in crevices in the mountain, it caused blocks of rock to break off to form Pulpit Rock.

Site status	**Geographic landmark**

Rock of Gibraltar

Natural wonders

Land marvels

Rocks

Factfile

Location	Europe – British territory, southern Spain
Part of	Gibraltar Nature Reserve
Size	1,398 feet high (maximum) by 2.6 square miles
Age	200 million years
Rock type	Shale, limestone and more
Known for	Its tunnel network, built by the British Army

Feature Gibraltar is famous for its 230 Barbary macaques. They are the only troop of wild monkeys in Europe and they have a reputation for stealing things.

Fact Gibraltar's rock layers are inverted – the youngest are at the bottom, the oldest at the top. The overturning could result from a tectonic plate collision.

Site status — **Nature reserve**

72

Seven Sisters

Factfile

Location	Europe – East Sussex, United Kingdom
Part of	South Downs National Park
Size	259 feet high (maximum) and 12 miles long
Age	84–87 million years
Rock type	Chalk
Known for	Their series of chalk cliffs

Feature Chalk is soft and it weathers. The cliffs are crumbling and washing away at the rate of 12–16 inches per year. Cliff falls occur 2–3 times a year.

Fact The bright-white, sheer cliffs of the Seven Sisters are an iconic UK landmark. Hawks and fulmars nest on narrow ledges on the cliffs.

Site status	National park

Devils Tower

Factfile

Location	North America – northeastern Wyoming, USA
Part of	Devil's Tower National Monument
Size	1,276 feet high and 800 feet wide at base
Age	50–60 million years
Rock type	Phonolite porphyry (an igneous rock)
Known for	Being sacred to the Lakota people

Feature This flat-topped rock is covered with hundreds of parallel, vertical cracks. It is a rock climbing site, and the fastest ascent has been 18 minutes!

Fact Devil's Tower is a sacred site for some 20 Native North American tribes. One ancient tribal story has it that a bear scored the tower with its claws.

Site status	National monument

74

Landscape Arch

Factfile

Location	North America – Moab, Utah, USA
Part of	Arches National Park
Size	290 feet long (base to base)
Age	150 million years
Rock type	Sandstone
Known for	Its colossal size

Feature Landscape Arch is believed to be the longest natural freestanding piece of rock in the whole world. The thinnest section is only six feet thick.

Fact Since the early 1990s, slabs – one measuring 73 feet – have fallen off the arch. The arch could collapse at any time and the path under it is now closed.

Site status	National park

Monument Valley

Natural wonders

Land marvels

Rocks

Factfile

Location	North America – Utah-Arizona border, USA
Part of	Monument Valley Navajo Tribal Park
Size	143 square miles (size of park)
Age	50 million years
Rock type	Sandstone
Known for	Its red sandstone buttes

Feature Giant buttes and pinnacles, 400–1,000 feet high, tower over the red desert floor. For many years, this area typified the American West of the movies.

Fact Water and wind have been leveling the ground and eroding the layers of rock to expose fascinating formations like the Mittens and Totem Pole.

Site status	Navajo Nation park

Pinnacles

Factfile

Location	North America – near Soledad, California, USA
Part of	Pinnacles National Park
Size	41 square miles (area of park)
Age	23 million years
Rock type	Rhyolite and andesite (volcanic rocks)
Known for	Being a release site for California condors

Feature The Pinnacles is an area of huge spiked rock formations that reach impressive heights of 1,200 feet. These rocks provide a landscape like no other.

Fact The Pinnacles were created by multiple volcano eruptions and movement on the San Andreas fault, shunting this section 195 miles northwest.

Site status   **National park**

Thor's Hammer

Natural wonders

Land marvels

Rocks

Factfile

Location	North America – southwestern Utah, USA
Part of	Bryce Canyon National Park
Height	150 feet
Age	40–60 million years
Rock type	Limestone
Known for	An unusual geological formation

Feature The top of this hoodoo (a thin rock pillar) is a giant hammerhead of rock. Erosion will cause the hammerhead to come crashing down one day.

Fact Thor's Hammer is named for the Norse god whose hammer created earthquakes. Bryce Canyon has more hoodoos than anywhere in the world.

Site status	National park

Paradise Bay

Factfile

Location	Antarctica
Part of	The Antarctic Peninsula
Environment	Glaciated mountains, ice cliffs and icebergs
Area	6 square miles (estimated)
Bay length	18 square miles (estimated)
Known for	Gentoo and chinstrap penguin colonies

Feature Also known as Paradise Harbor, this wide bay is edged with glaciated mountains and ice cliffs, which are reflected in the calm water on a sunny day.

Fact Paradise Bay is one of Antarctica's most visited spots, with cruise ships stopping to let passengers admire the view and walk among the penguins.

Site status   **Nature reserve**

Ha Long Bay

Natural wonders

Water treasures

Bays

Factfile

Location	Asia – Quang Ninh Province, Vietnam
Part of	Gulf of Tonkin
Environment	Stone islands, grottoes and caves
Area	579 square miles
Bay length	75-mile-long coastline
Known for	Limestone islands, some hollowed out by caves

Feature There are about 1,600 islands in Ha Long Bay. Being steep and conical, most of the islands are uninhabited. Fishermen live on floating villages in the bay.

Fact Vietnamese legend says that the bay was created by a dragon, its tail gouging out the earth and creating all the islands as it jumped into the water.

Site status World Heritage Site

80

Phang Nga Bay

Factfile

Location   Asia – Phang Nga Province, Thailand
Part of   Ao Phang Nga National Park
Environment   Limestone islets, caves, reefs and mangroves
Area   154 square miles
Islet height   1,150 feet (maximum)
Known for   James Bond Island that featured in a Bond film

Feature Phang Nga Bay has 42 gravity-defying islets that jut out of the emerald-green water. Wave action is eroding the stacks by three feet every 5,000 years.

Fact The water in Phang Nga Bay is only a few yards deep and kayaking is the best way to explore the grottoes, caves and archaeological sites.

Site status   **National park**

Bay of Islands

Factfile

Location	Oceania – North Island, New Zealand
Part of	Bay of Islands Maritime Park
Environment	A natural harbor with beaches and islands
Area	100 square miles
No. of islands	144
Known for	Hole in the Rock on Piercy Island (Motu Kokako)

Feature With a subtropical climate, the Bay of Islands, which is a drowned valley, is famous for its beauty and its big-game fishing, sailing and diving.

Fact The Bay of Islands has historical significance. The first Maori arrived here 700 years ago, and in 1814 it became the first British settlement.

Site status	Regional park

82

Bay of Fundy

Factfile

Location North America – Atlantic coast, Canada
Part of Bay of Fundy UNESCO Biosphere Reserve
Environment Rock cliffs and rock formations, and forests
Area 1,732 square miles
Bay length 170 miles
Known for The world's highest tides reaching 53.5 feet

Feature The steep rocky cliffs form two narrow channels that funnel 160 billion tons of seawater twice a day to create the bay's phenomenal high tides.

Fact The extreme tides of the Bay of Fundy have carved sea cliffs, sea stacks and caves into the sandstone, while its volcanic headlands resist the tides.

Site status	Nature reserve

The Great Geysir

Factfile

Location	Europe – Haukadalur Valley, Iceland
Part of	Geysir Geothermal Field
Height	200 feet
Activity	Infrequent but active for 10,000 years
Temperature	257 °F (65 feet down the vent)
Known for	The first geyser known to modern Europeans

Feature The word "geyser" comes from a misspelling of Geysir. Geysers are vents in volcanically active areas that spout naturally heated water.

Fact Geysir is most active after an earthquake. In the 1980s people were too impatient to wait, so they dropped soap into the geyser to make it spout.

Site status   **National park**

Castle Geysir

Factfile

Location	North America – Wyoming, USA
Part of	Yellowstone National Park
Height	60–90 feet
Activity	Every 11–14 hours
Temperature	200 °F
Known for	Its 12-foot-tall castle-like cone

Feature The column of hot water erupts for about 20 minutes and is followed by a 30–40-minute steam phase. The cone has been carbon-dated to 1022.

Fact Yellowstone has over 300 geysers and some 10,000 hydrothermal features in total, which means it is home to half of the world's geothermal wonders.

Site status	World Heritage Site

85

Fly Geyser

Factfile

Location   North America – Washoe County, Nevada, USA
Part of   Fly Ranch
Height   5 feet
Activity   Constant since 1964
Temperature   200 °F
Known for   Being rainbow colored with multiple spouts

Feature A small mound, only five feet high, the Fly Geyser's colorful cone is covered with green and red algae that thrive in the hot, moist conditions.

Fact When miners were drilling a well, they hit geothermal water. The minerals in the water created Fly Geyser's "cone," which is still growing today.

Site status   **Private land**

El Tatio Geysers

Factfile

Location	South America – San Pedro de Atacama, Chile
Part of	El Tatio Geyser Nature Reserve
Height	29 inches (average) but can reach 19 feet
Activity	Daily
Temperature	185 °F
Known for	Being the world's third-largest geyser field

Feature El Tatio, meaning "oven" in Quechuan, is a geyser field with 40 active geysers, 62 hot springs, 85 fumaroles and five mud volcanoes.

Fact At an elevation of 13,000 feet in the Andes Mountains, the best time to witness the geysers at their best is between 6 a.m. and 7 a.m.

Site status	**Nature reserve**

Fox Glacier

Factfile

Location	Oceania – Weheka, South Island, New Zealand
Part of	Te Wahipounamu World Heritage Site
Size of park	279 miles long by 24–55 miles wide
Length	8.1 miles
Height	9,500 feet
Known for	Its descent into a lush temperate rainforest

Feature The longest of New Zealand's West Coast glaciers and one of the world's most accessible to visitors, the Fox Glacier is an extraordinary vista to behold.

Fact Due to its shape, the Fox Glacier moves much faster than most valley glaciers. The ice slides down a steep descent of 8,500 feet to a river valley.

Site status   World Heritage Site

Margerie Glacier

Factfile

Location	North America – Glacier Bay, Alaska, USA
Part of	Glacier Bay National Park and Preserve
Size of park	5,130 square miles
Length	21 miles
Height	100 feet below sea level and 250 feet above
Known for	Producing icebergs via the calving process

Feature Beginning at Mount Root, the one-mile-wide Margerie Glacier descends to the sea. This pristine blue-ice glacier is reached by boat or plane.

Fact This tidewater glacier – starting on land and ending in seawater – "calves," which is when walls of ice crack off the glacier and thunder into the sea.

Site status   **National park**

Perito Moreno Glacier

Natural wonders

Water treasures

Glacier

Factfile

Location	South America – Santa Cruz Province, Argentina
Part of	Los Glaciares National Park
Area of park	2,317 square miles
Length	19 miles
Height	4,921 feet
Known for	Its process of fracturing to form icebergs

Feature This glacier is still growing. It originated on the Ice Caps (a vast continental ice extension) and descends 4,000 feet to end at the edge of Argentino Lake.

Fact When the front of the 196-foot-thick glacier blocks Lake Argentino, the water can rise 65 feet and the glacier fractures to make icebergs.

Site status   **World Heritage Site**

Lake Nakuru

Factfile

Location	Africa – Great Rift Valley, Kenya
Area	2–17 square miles (depending on the season)
Depth	10 feet (maximum)
No. of Islands	0
Water type	Saline
Known for	Its flocks of flamingos

Feature The Lake Nakuru National Park was enlarged to make a poacher-proof sanctuary for Rothschild's giraffes and black and white rhinos.

Fact Nakuru is home to 75 percent of the world's lesser flamingos. They feed on algae created by lake water, plankton and bird droppings!

Site status   World Heritage Site

Lake Baikal

Factfile

Location   Europe – Siberia, Russia

Area   12,200 square miles

Depth   5,315 feet (maximum)

No. of islands 45

Water type   Freshwater

Known for   Being the world's oldest and deepest lake

Feature Known as the "Galapagos of Russia," many unusual animals, like the wolverine, brown bear, Eurasian lynx and reindeer, are found here.

Fact This record-breaking lake is 25 million years old, it contains 20 percent of Earth's unfrozen freshwater and is one of the world's clearest lakes!

Site status   **World Heritage Site**

Plitvice Lakes

Factfile

Location	Europe – County of Lika-Senj, Croatia
Area	0.77 square miles
Depth	3–154 feet
No. of lakes	16
Water type	Freshwater
Known for	Its series of waterfalls, one dropping 230 feet

Feature The water in the lakes varies in color – from azure blue to green, blue and gray. This is due to the level of minerals and organisms in the water.

Fact Sixteen lakes, linked by cascades, create the Plitvice Lakes. The lakes are separated by naturally formed travertine (a form of limestone) dams.

Site status   **World Heritage Site**

Great Salt Lake

Factfile

Location	North America – Utah, USA
Area	1,700 square miles
Depth	16 feet (average)
No. of Islands	11 (varies depending on water level of lake)
Water type	Saline
Known for	Being a very salty, closed (no outflow) lake

Feature One of the largest salt flats near the Great Salt Lakes is Bonneville. It is the fastest race track in the world and home to many land speed records.

Fact Though fed by three freshwater rivers, the minerals in the lake bed make the water saltier than seawater. The lake is too salty for most aquatic species.

Site status   **Regional park**

94

Moraine Lake

Factfile

Location	North America – Alberta, Canada
Area	0.19 square miles
Depth	45 feet (maximum)
No. of Islands	0
Water type	Freshwater
Known for	Its scenic setting in the Valley of Ten Peaks

Feature Fed by a glacier, Lake Moraine is famous for its sparkling blue waters, ice-capped peaks and lush, green forests. It is at its best from mid-June–end July.

Fact You have to be "bear aware" at Lake Moraine. It is a very important habitat for grizzly bears and is where females feed and raise their cubs.

Site status   **National park**

95

Spotted Lake

Natural wonders

Water treasures

Lakes

Factfile

Location	North America – British Columbia, Canada
Area	0.06 square miles
No. of pools	365 (approximate)
No. of islands	0
Water type	Mineral / saline
Known for	Lake evaporating leaving spots of water

Feature In the summer when the water level drops, shallow pools form. When these evaporate, "spots" of white, yellow, green or blue minerals are left behind.

Fact The dried spots of minerals, which includes magnesium sulfate, calcium, titanium and silver, used to be gathered to make ammunition.

Site status	**Nature reserve**

Lake Titicaca

Factfile

Location   South America – Andes Mountains, Peru/Bolivia
Area   3,232 square miles
Depth   932 feet (maximum)
No. of islands 5 plus 44 artificial floating islands
Water type   Freshwater
Known for   Being the world's highest navigable lake

Feature Lake Titicaca is fed by 27 glacier-water rivers and drains into one small river. Serious water pollution saw this lake given threatened status in 2012.

Fact Legend says that the mythical founder of the Inca civilization was released from the lake by Inti, the sun god, in order to establish the Incan Empire.

Site status	Nature reserve

Salar de Uyuni

Factfile

Location	South America – Andes Mountains, Bolivia
Area	4,247 square miles
Depth	4–8-inch salt crust with 400 feet salt below
No. of islands	33 rock outcrops often covered with cactus
Water type	Salt crust with salt/mineral/water (brine) below
Known for	Being the world's largest salt flat

Feature The lake's upper layer is a flat, ice-white dazzling crust of hard salt that can support the weight of a car. Locals build houses using bricks of salt.

Fact The lake was part of Lago Minchin, a prehistoric salt lake that covered southwest Bolivia. Salar de Uyuni contains 10 billion tons of salt.

Site status   **National park**

Mesoamerican Barrier Reef

Factfile

Location   North America – Caribbean Sea
Length   700 miles
Depth   25–125 feet
No. of islands 3 offshore atolls and several hundred sand cays
Water type   Salt
Known for   The second longest barrier reef in the world

Feature The reef is home to many fish, turtle, shark and coral species, but it is threatened by overfishing, rising water temperature and pollution.

Fact The Mesoamerican Reef is sometimes within yards of the shore, creating a special "mix" of reef, seagrass and mangrove habitats and species.

Site status	World Heritage Site

Raja Ampat Reef

Factfile

Location   Asia – part of the Coral Triangle, West Papua
Area   15,444 square miles
Depth   32–131 feet
No. of islands 4 islands and over 1,500 islets, cays and shoals
Water type   Salt
Known for   Being home to a phenomenal number of species

Feature These delicate coral reefs are nourished by the powerful deep-sea currents. It is one of the top scuba diving destinations in the world.

Fact During the first major scientific study in 2001, over 1,300 fish species were discovered around the reef. Many species had never been seen before.

Site status   **National park**

The Great Barrier Reef

Factfile

Location	Australia – east coast Queensland
Length	132,973 square miles
Depth	114–6,561 feet
No. of islands	900 islands and cays and 150 mangrove islands
Water type	Salt
Known for	Being the largest living thing on Earth

Feature Over 2,500 coral reefs make up the Great Barrier Reef. It is home to 400 types of coral, 1,625 species of fish, 133 ray and shark species, and more.

Fact The reef is under threat from over-fishing, tourism and commercial shipping and the crown of thorns starfish that eats the living coral.

Site status — **World Heritage Site**

The Lagoons of New Caledonia

Natural wonders

Water treasures

Reefs

Factfile

Location   Oceania – southwest Pacific Ocean
Area   9,035 square miles
Depth   82 feet (average)
No. of islands Includes several large and small islands
Water type   Salt
Known for   The discovery of rare and new marine species

Feature The reef that surrounds the lagoon is actually two barrier reefs and is one of the three most extensive reef systems in the world.

Fact The Lagoons of New Caledonia are home to the world's third-largest population of dugongs. The area is also the nesting ground for green sea turtles.

Site status   **World Heritage Site**

102

Red Sea Coral Reef

Factfile

Location	Africa/Asia – bounded by 6 countries
Length	1,240 miles
Temperature	93 °F (summer) / 82 °F (winter)
No. of islands	Several islands and atolls
Water type	Salt
Known for	Exceptional water clarity

Feature The Red Sea reef has an unusually high tolerance of high temperatures and high levels of salt, which would destroy most other coral reefs.

Fact The coral reef formations are believed to have been caused as a result of tectonic forces taking place over the last million years.

Site status **National park**

Niger River

Factfile

Location	Africa – through 9 West African countries
Length	2,600 miles
Width	2.4 miles (in flood)
Drainage area	730,000 square miles (7.5 percent of Africa)
Drains into	Gulf of Guinea, Atlantic Ocean
Known for	Inner Niger Delta is the size of Belgium

Feature This river starts just 150 miles from the Atlantic Ocean, but then flows away from the sea into the Sahara before heading southeast, back to the Atlantic.

Fact The third-longest river in Africa and a vital source of water to many tribes, the Niger passes close to Timbuktu, the most remote city in the world.

Site status	Major world river

The Nile

Factfile

Location	Africa – through 11 northeast African countries
Length	4,132 miles (estimates vary)
Width	1,148 feet–7.5 miles
Drainage area	1,256,591 square miles (10 percent of Africa)
Drains into	Mediterranean Sea
Known for	Being the longest river in the world

Feature There are many reservoirs and dams along the Nile, but this has not prevented millions of people along its length being affected by droughts.

Fact The Nile is strongly associated with ancient Egyptians, but in reality only 22 percent of the Nile's course runs through Egypt's fertile valley.

Site status	**Major world river**

Mackenzie River

Factfile

Location   North America – Northwest Territories, Canada
Length   1,079 miles
Width   0.3–3.1 miles
Drainage area   697,000 square miles (20 percent of Canada)
Drains into   Beaufort Sea, Arctic Ocean
Known for   Being part of Canada's longest river system

Feature The Mackenzie is a slow-moving river and from start to end forms a stunning network of streams, lakes, channels, sandbars and islands.

Fact During the winter months, it is hard to see the Mackenzie River at all as it freezes over. Some sections of the frozen river are used as ice roads.

Site status	Major world river

106

Mississippi River

Factfile

Location	North America – through 10 states of the USA
Length	2,350 miles
Width	20 feet–7 miles
Drainage area	1.2 million square miles (40 percent of 48 states)
Drains into	Gulf of Mexico
Known for	Being the fourth longest river in the world

Feature The famous steamboats entered trade in the 1820s to transport cotton, timber and food down the river. A few steamboats still use the river.

Fact In 2002, Martin Strel swam the length of the Mississippi in 68 days. He has also swum the Amazon, Yangtze, Danube and Paraná rivers.

Site status   **Major world river**

Amazon River

Factfile

Location	South America – through 7 countries
Length	3,903–4,195 miles
Width	6.8 miles (dry season)–120 miles (rainy season)
Drainage area	2.7 million square miles (40 percent of South America)
Drains into	Atlantic Ocean
Known for	Carrying the greatest volume of water

Feature There is not a single bridge across the Amazon. This is because it runs through rain forests, not cities. Ferries make the crossings instead.

Fact When the Amazon reaches the Atlantic, its estuary is 205 miles wide and deep enough for small ocean ships to travel inland for about 2,300 miles.

Site status	Major world river

Red Sea

Factfile

Location	Africa/Asia – surrounded by 9 countries
Surface area	169,100 square miles
Length	1,200 miles
Width	190 miles (maximum)
Depth	1,640 feet (average) / 8,200 feet (maximum)
Known for	Being the world's northernmost tropical sea

Feature Low rainfall, high temperatures, high rates of evaporation and salty water from the gulfs of Aden and Suez mean that the Red Sea is very salty.

Fact The Red Sea most likely got its name from an orange-red algae that blooms seasonally on the surface of the clear blue-green water.

Site status	Marginal sea

Dead Sea

Factfile

Location	Asia – bordered by 2 countries and a territory
Surface area	394 square miles
Length	50 miles
Width	11 miles
Depth	10 feet (minimum)–1,300 feet (maximum)
Known for	Being the lowest place on Earth

Feature The Dead Sea is 1,407 feet below sea level in a graben, a block of Earth's crust that has been forced down by collision with another section of crust.

Fact The Dead Sea is the deepest hypersaline (9.6 times as salty as the ocean) lake in the world. With its high density, floating is made easy in the Dead Sea.

Site status	**Marginal sea**

Black Sea

Factfile

Location	Europe/Asia – bounded by 6 countries
Surface area	163,000 square miles
Length	730 miles
Width	160 miles
Depth	7,250 feet (maximum)
Known for	Being on the "crossroads" of East meets West

Feature Though linked to the Mediterranean Sea (saltwater), the Black Sea is freshwater, does not have tides and the water is always flat, calm and quiet.

Fact Below 650 feet no life can survive in the Black Sea. This is because the deep water lacks oxygen and contains a zone of poisonous "rotten egg" gas.

Site status	**Marginal sea**

Caspian Sea

Factfile

Location	Europe/Asia – bounded by 5 countries
Surface area	143,000 square miles
Length	750 miles
Width	200 miles
Depth	3,360 feet (maximum)
Known for	Being the world's largest inland body of water

Feature This saltwater sea is primarily freshwater where the Volga – the largest river in Europe – enters it. The Caspian is famous for its caviar-producing sturgeon fish.

Fact The Caspian Sea was part of an enormous prehistoric sea. When the land rose and water level fell 5.5 million years ago, the Caspian became landlocked.

Site status	Marginal sea

Sargasso Sea

Factfile

Location	North Atlantic Ocean
Surface area	1.6 million square miles (estimate)
Length	1,988 miles
Width	683 miles
Depth	23,000 feet (maximum)
Known for	Being the only sea without land boundaries

Feature This sea is known for its sargassum, a free-floating seaweed that provides shelter and food for many marine species, like turtles, crabs, eels and shrimp.

Fact The Sargasso Sea is located entirely within the Atlantic Ocean. It is defined only by ocean currents, like the Gulf Stream, that surround it.

Site status	Non-marginal sea

Seljalandsfoss Waterfall

Factfile

Location	Europe – South Region, Iceland
Part of	Seljalandsfoss River
Height	200 feet
Width	82 feet (estimated)
No. of drops	1
Known for	The walkable trail that goes behind the waterfall

Feature The Seljalandsfoss – "foss" is Icelandic for "waterfall" – is unusual in that the water falls from the cliff in a thin curtain straight into a deep pool.

Fact The path that runs behind the base of the waterfall is open during the summer, and on sunny days a double rainbow often forms around the falls.

Site status	Regional park

Niagara Falls

Natural wonders

Water treasures

Waterfalls

Factfile

Location	North America – Canada/USA border
Part of	Niagara River
Height	70–188 feet
Width	3,409 feet (total width)
No. of drops	3
Known for	The volume of water that flows over the falls

Feature Niagara Falls straddles the border between Ontario and New York State. There are three waterfalls: Horseshoe, American and Bridal Veil.

Fact About six million cubic feet of water crashes over the top of the falls every couple of minutes and it falls at a speed of 68 miles per hour.

Site status   **National park**

Angel Falls

Factfile

Location	South America – southeastern Venezuela
Part of	Churún River
Height	3,212 feet
Width	500 feet (at base)
No. of drops	2
Known for	Being the highest waterfall in the world

Feature The falls spill from a flat-topped mountain hardly touching the sheer cliff. It is said that drops of water spread up to a mile away from these waterfalls.

Fact This waterfall was named after James Angel, who crash-landed his plane above the falls. It took Angel 11 days to climb down the falls' cliff face.

Site status   **Conservation area**

Iguazu Falls

Factfile

Location	South America – Argentina/Brazil border
Part of	Iguazu River
Height	262 feet (maximum)
Width	1.7 miles
No. of drops	150–300 (depending on the season)
Known for	The sprays of water from the many cascades

Feature This horseshoe-shaped waterfall, its name meaning "great water," produces so much spray that it drenches the surrounding forests.

Fact The large number of mini-waterfalls are created by islands on the Iguazu River that split the flow of water. Some parts of the falls never cascade water.

Site status   World Heritage Site

Aurora

Factfile

Where	Magnetic north and south poles
What	Vivid colors taking many forms in the night sky
When	During times of high solar activity
Duration	Varies
Caused by	Particles from the Sun hitting Earth's gases
Color	Predominantly green

Feature The lights are caused by electrically charged particles from the Sun colliding with gases, like oxygen and others, in Earth's atmosphere.

Fact The Aurora at the North Pole is the Borealis, and that at the South Pole is the Australis. The lights appear 50–400 miles above Earth's surface.

Site status	Atmospheric optics

Midnight Sun

Factfile

Where	Within the Arctic and Antarctic circles
What	When the Sun remains visible for 24 hours a day
When	During summer months
Duration	Up to six months
Caused by	The 23-degree tilt of Earth's axis
Color	Normal sunlight

Feature Few see a midnight sun in Antarctica, but an Arctic one is seen in parts of Canada, Greenland, Iceland, Finland, Norway, Russia and Alaska.

Fact On "white nights" – nights that never get completely dark – beyond the polar circles, the sun sets and then immediately rises above the horizon.

Site status	Atmospheric optics

Blue Grotto

Factfile

Where	Europe – Biševo Island, Croatia
What	Sunlight illuminates a cave to glow blue
When	11 a.m.–noon on a sunny day (varies with season)
Duration	Variable
Caused by	Reflection of sunlight on cave floor
Color	Aquamarine

Feature The Blue Grotto is not the only natural light show in the area. Not far away is the larger Green Grotto, which is bathed in an emerald-green glow.

Fact The glowing blue is caused by the Sun's rays entering the sea cave through the water and reflecting off the cave's white limestone floor.

Site status   **Atmospheric optics**

Horsetail Fall

Factfile

Where	North America – Yosemite National Park, USA
What	The falls appear to be red orange in color
When	Middle to late February, during sunset
Duration	10 minutes
Caused by	A setting sun striking the falls at a certain angle
Color	Red orange

Feature At those times in the year when the Sun sets at a particular angle, it illuminates the waterfall red orange to resemble a ribbon of fire.

Fact The Yosemite fire fall, as it is known, is very temperamental. It glows only when conditions, like temperature and weather, are absolutely perfect.

Site status   **Atmospheric optics**

122

Light Pillars

Factfile

Where	Most frequent in polar regions
What	A column of light appearing to beam upwards
When	During cold conditions
Duration	Varies
Caused by	Light reflecting off ice crystals
Color	Varies depending on the source of light

Feature Light pillars are optical illusions, but are often thought to be UFOs. Pillars occur in winter around Niagara Falls when the spotlights fall on ice crystals.

Fact Light pillars, or haloes, are caused when light reflects off flat ice crystals in the air near Earth's surface. The crystals act like a mirror.

Site status   **Atmospheric optics**

Parhelic Circle

Factfile

Where	Can occur anywhere on Earth
What	A horizontal line across the sky at Sun height
When	A sunny day with the correct conditions
Duration	Variable
Caused by	Light reflecting off ice crystals
Color	Generally white

Feature The photo above shows the parhelic circle across the face of the Sun, a halo around the Sun, two "sundogs" on the halo and an arc halo at the top.

Fact This rare optical phenomenon occurs only when light from the Sun reflects off near-vertical faces of ice crystals. It can also occur with moonlight.

Site status	Atmospheric optics

124

Solar eclipse

Factfile

Where	Visible from anywhere on Earth
What	The Sun's corona appears behind the Moon
When	Approximately every 18 months
Duration	Up to 7.5 minutes
Caused by	The Moon passing between the Sun and Earth
Color	White glow

Feature During a solar eclipse, the Moon appears the same size as the Sun, day turns to night and the Sun's chromosphere (one of its gas layers) can be seen.

Fact A solar eclipse occurs when a new Moon is directly aligned between the Sun and Earth. The Moon will cast its shadow on Earth's surface.

Site status   **Atmospheric optics**

Glossary

Alkaline A substance that is not acidic.

Alluvial Loose sediment that was deposited by a river.

Archipelago A chain or cluster of islands.

Avalanche A mass of ice, snow and rocks falling down a mountainside.

Basalt A rock formed from lava.

Badland Eroded land that has little vegetation.

Bedrock Rocks that lie under loose soil.

Butte An isolated hill that juts sharply up from the land.

Canopy The uppermost branches and leaves of a tree.

Canyon A deep split in the land, often with a river flowing through it.

Chalk A soft white form of limestone.

Chamber/cavern Other words to describe caves.

Composite rock A rock changed into another rock by heat or pressure.

Desertification The process that turns land into desert.

Dolomite A rock made by sediments compressed under water.

Dormant When a volcano is inactive for a period.

Dripstone A rock formed by dripping water that contains salts and minerals.

Drowned valley A valley that is flooded after a rise in water level.

Erosion The movement of loose soil and rock fragments by wind or water.

Estuary Where a river meets a tidal body of water.

Fjord A deep but narrow piece of water between cliffs or mountains.

Fumaroles A vent near a volcano that emits gases.

Geothermal Heat produced by processes inside Earth.

Glacier A large, slow-moving mass of ice and rock.

Gorge A narrow valley between mountains, often with a river flowing along it.

Granite A hard igneous rock.